# hands-on maths

## Kerry Dalton

# Year 2

Published by Keen Kite Books
An imprint of HarperCollins*Publishers* Ltd
The News Building
1 London Bridge Street
London
SE1 9GF

HarperCollins*Publishers*
1st Floor, Watermarque Building, Ringsend Road
Dublin 4, Ireland

ISBN 9780008266967

First published in 2017

10 9 8 7

Series Concept and Commissioning: Shelley Teasdale and Michelle I'Anson
Project Manager: Fiona Watson
Editor: Denise Moulton
Cover Design: Anthony Godber
Text Design and Layout: Contentra Technologies
Production: Natalia Rebow
Printed and bound in the UK using 100% Renewable
Electricity at CPI Group (UK) Ltd

A CIP record of this book is available from the British Library.

# Contents

# Year 2 aims and objectives

*Hands on Maths Year 2* encourages pupils to enjoy a range of mathematical concepts through a practical and hands-on approach. Using a range of everyday objects and mathematical resources, pupils will explore and represent key mathematical concepts. These concepts are linked directly to the National Curriculum 2014 objectives for Year 2. Each objective will be investigated over the course of a week, using a wide range of hands-on approaches such as cubes, 100 squares, dominoes, toy animals, counters and cards. The mathematical concepts are explored in a variety of contexts to give pupils a richer and deeper learning experience, which enables mastery to be attained.

## Year 2 programme and overview of objectives

| Topic | Week 1 | Week 2 | Week 3 | Week 4 | Week 5 | Week 6 |
|---|---|---|---|---|---|---|
| **Counting** | Count in steps of 2 from 0, forward and backward | Count in steps of 3 from 0, forward and backward | Count in steps of 5 from 0, forward and backward | Count in tens from any number, forward and backward | Count in tens from any number, forward and backward | Count in steps of 2, 3 and 5 from 0, and in tens from any number, forward and backward |
| **Place value** | Recognise the place value of each digit in a two-digit number (tens and ones) | Recognise the place value of each digit in a two-digit number (tens and ones) | Read and write numbers to at least 100 in numerals and in words | Compare and order numbers from 0 up to 100; use = sign | Compare and order numbers from 0 up to 100; use < sign | Compare and order numbers from 0 up to 100; use > sign |
| **Representing numbers** | Identify and represent numbers 0–100 using different representations | Identify and represent numbers 0–100 using different representations | Identify and represent numbers 0–100 using number lines | Estimate numbers 0–50 | Estimate numbers 0–100 | Compare and order numbers up to 100; use <, > and = signs |
| **Addition and subtraction** | Recall and use addition and subtraction facts to 20 fluently | Recall and use addition and subtraction facts to 20 fluently, and derive and use related facts up to 100 | Add and subtract a two-digit number and ones | Add and subtract a two-digit number and tens | Add and subtract two two-digit numbers | Add three one-digit numbers |

# Year 2 aims and objectives

| Topic | Week 1 | Week 2 | Week 3 | Week 4 | Week 5 | Week 6 |
|---|---|---|---|---|---|---|
| **Multiplication and division** | Recall and use multiplication and division facts for the 2 multiplication table | Recall and use multiplication and division facts for the 2 multiplication table, including recognising odd and even numbers | Recall and use multiplication and division facts for the 5 multiplication table | Recall and use multiplication and division facts for the 10 multiplication table | Recall and use multiplication and division facts for the 2, 5 and 10 multiplication tables | Show that multiplication of two numbers can be done in any order (commutative) and division of one number by another cannot |
| **Fractions** | Recognise, find, name and write fractions $\frac{1}{2}$ of a length, shape, set of objects or quantity | Recognise, find, name and write fractions $\frac{1}{4}$ of a length, shape, set of objects or quantity | Recognise, find, name and write fractions $\frac{2}{4}$ of a length, shape, set of objects or quantity | Recognise, find, name and write fractions $\frac{3}{4}$ of a length, shape, set of objects or quantity | Recognise, find, name and write fractions $\frac{1}{3}$ of a length, shape, set of objects or quantity | Recognise, find, name and write fractions $\frac{1}{3}$, $\frac{1}{4}$, $\frac{2}{4}$ and $\frac{3}{4}$ |

# Introduction

The *Hands-on maths* series of books aims to develop the use of readily available manipulatives such as toy cars, shells and counters to support understanding in maths. The series supports a concrete–pictorial–abstract approach to help develop pupils' mastery of key National Curriculum objectives.

Each title covers six topic areas from the National Curriculum (counting, representing numbers, understanding place value, the four number operations: addition and subtraction and multiplication and division, and fractions). Each area is covered during a six-week unit, with an easy-to-implement 10-minute activity provided for each day of the week. Photos are included for each activity to support delivery.

*Hands-on maths* enables a deep interrogation of the curriculum objectives, using a broad range of approaches and resources. It is not intended that schools purchase additional or specialist equipment to deliver the sessions; in fact, it is hoped that pupils will very much help to prepare resources for the different units, using a range of natural, formal and typical maths resources found in most classrooms and schools. This will help pupils to find ways to independently gain a deep understanding and enjoyment of maths.

A typical 'hands-on' classroom will have a good range of resources, both formal and informal. These may include counters, playing cards, coins, Dienes, dominoes, small objects such as toy cars and animals, Cuisenaire rods, hundred squares and hoops.

There is no requirement to use *only* the resources seen in the photographs that accompany each activity. Cubes may look like those in the green bowl, or will be just as effective if they look like the ones in the blue bowl. They serve the same purpose in helping pupils understand what the cubes represent.

# Resources

*Hands-on maths* uses a range of formal, informal and 'typical' resources found in most classrooms and schools. To complete the activities in this book, it is expected that teachers will have the following resources readily available:

- whiteboards and pens for individual pupils and pairs
- Dienes and Cuisenaire rods
- dice, coins and bead-strings
- a range of cards, including playing cards, place-value arrow cards and numeral cards

- collections of objects that pupils are interested in and want to count, such as toy cars, toy animals and shells
- bowls/containers to store sets of resources in, making it easy for pupils to handle and use the objects

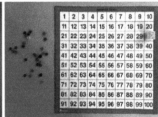

- ten frames (these could be egg boxes, ice-cube trays, printed frames or something pupils have created themselves)
- number lines and hundred squares – lots of different types and styles: printed, home-made, interactive, digital or practical … whatever you prefer, and whatever is handy. (For hundred squares, there is, of course, the 1–100 or 0–99 choice to make; both work and it is best to choose whatever works for the class. Both offer a slight difference in place value perspective, with 0–99 giving the 'zero as a place holder' emphasis, while the 1–100 version helps pupils to visualise the position of 100 in relation to the two-digit numbers.)

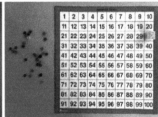

- counters and cubes – lots of them! Many of the activities require counters and cubes to be readily available. The cubes can be any size and any colour: what the cubes represent is the most important factor.

Maths is a truly unique, creative and exciting discipline that can provide pupils with the oppportunity to delve deeply into core concepts. Maths is found all around us, every day, in many different forms. It complements the principles of science, technology and engineering.

*Hands-on maths* provides ideas that can be adapted to suit the broad range of needs in our classrooms today. These ideas can be used as a starting point for assessment – before, during or after teaching a particular topic has taken place. The activities are intended to be flexible enough to be used with a whole class and can, of course, be differentiated to suit individual pupils in a class.

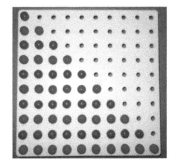

The activities can be adapted to link to other subject areas and interests. For example, a suggestion to use farm animals may link well to a science unit on classification or food chains; alternatively, the resource could be substituted with bugs if minibeasts is an area of interest for pupils. Teachers can be as flexible as they wish with the activities and resources – class teachers know their pupils best.

Spoken language is underpinned in maths by the unique mathematical vocabulary pupils need to be able to use fluently in order to demonstrate their reasoning skills and show mathematical proof. The correct, regular and secure use of mathematical language is key to pupils' understanding; it is the way in which they reason verbally, negotiate conceptual understanding and build secure foundations for a love of mathematics and all that it brings. Each unit in *Hands-on maths* identifies a range of vocabulary that is typical, but by no means limited to, that particular unit. The way the vocabulary is used and incorporated into activities is down to individual style and preference and, as with all of the resources in the book, will be very much dependent on the needs of each individual class. A blank template for creating vocabulary cards is included at the back of this book.

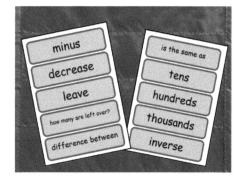

# Week 1: Counting

## Count in steps of 2 from 0, forward and backward

**Resources:** objects, bowls, cubes

**Vocabulary:** number, numeral, zero, one, two, three …, twenty, thirty …, one hundred, none, how many?, count, count up / on / down / back, count in ones, twos, threes, fives, tens, many, few, odd, even, every other, how many times?, pattern, pair, ones, tens, exchange, digit, 'teen' numbers, first, second, third … hundredth, last, before, after, next, between, multiple of, sequence, continue, predict

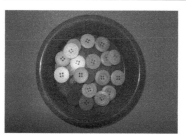

### Monday

Give each pupil a bowl containing at least 20 objects.

Each pupil counts out 20 objects. First, ask pupils to place their objects in a single row. Next, count together from 0–20 in ones, pointing to each object in turn.

Then ask pupils to put the objects into pairs. Count out loud together, saying the odd numbers quietly and the even numbers (i.e. the multiples of two) loudly.

### Tuesday

Give each pupil 20 cubes, 10 of one colour and 10 of a different colour.

Pupils create a tower of 20 cubes, alternating the colours. Ask pupils to count together from 0–20 in ones, and then count backwards to 0. Next, ask pupils to skip count by counting every other one; start at 0 (0, 2, 4, 6, 8 … 20), and then count backwards to 0.

### Wednesday

Use Tuesday's cubes.

Count together from 0–20 in twos, and then count backwards in twos to 0. Ask pupils to sort the cubes into twos and place next to each other in a column. Together, count the cubes again; start at the bottom and move up the column counting in twos to 20, then count backwards to 0.

### Thursday

Pupils work in fours to create a tower of 40 cubes, using the cube towers from earlier in the week.

Ask pupils to count from 0–40 in ones, and then count backwards to 0. Next, ask them to skip count by counting every other one; start at 0 (0, 2, 4, 6, 8 … 40), and then count backwards to 0.

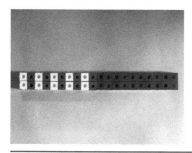

### Friday

Using Thursday's towers of 40 cubes and working in pairs, pupils count from 0–40 in twos, and then backwards to 0 in twos. Ask pupils to sort the cubes into twos and place next to each other in a column. They count the cubes again, starting at the bottom and moving up the column counting in twos. When they reach 40, they count backwards to 0.

## Count in steps of 3 from 0, forward and backward

**Resources:** objects, cubes, 100 squares

**Vocabulary:** number, numeral, zero, one, two, three …, twenty, thirty …, one hundred, none, how many?, down / back, count in ones, twos, threes, fives, tens, many, few, odd, even, every other, how many times?, pattern, pair, ones, tens, exchange, digit, 'teen' numbers, first, second, third … hundredth, last, before, after, next, between, multiple of, sequence, continue, predict

---

### Monday

Give each pupil 30 objects.

Together count from 0–30 in ones, pointing to each object in turn.

Ask the pupils to put the objects into threes. Count out loud together, saying the multiples of three loudly and the other numbers quietly.

---

### Tuesday

Give each pupil 30 cubes, 10 each of three different colours, placed as shown.

First, count together from 0–30 in ones, and then count backwards to 0. Next, ask pupils to skip count by counting every third one; start at 0 (0, 3, 6, 9, 12 … 30) and then count backwards to 0.

---

### Wednesday

Use Tuesday's cubes.

Count together from 0–30 in threes, and then count backwards in threes to 0. Highlight that pupils count aloud on every black cube. Ask pupils to separate the cubes into threes and place in a column. Together, count the cubes again; start at the bottom and move up the column counting in threes to 30. Then count backwards to 0.

---

### Thursday

Display a large 100 square.

Count quietly from 0–30 in ones. When you reach a multiple of three, encourage pupils to shout out the multiples of three while you circle the number. When you reach 30, count backwards to 0 in the same way.

Ask if pupils can see a pattern when counting in steps of 3.

---

### Friday

Display a large 100 square.

Remind pupils about Thursday's activity. Ask what patterns pupils remember when counting in steps of 3.

Count quietly from 0–60 in ones. When you reach a multiple of three, circle the number on the board and encourage pupils to shout out the multiples of three. When you reach 60, count backwards to 0 in same way.

---

# Week 3: Counting

## Count in steps of 5 from 0, forward and backward

**Resources:** cubes, 100 squares

**Vocabulary:** number, numeral, zero, one, two, three …, twenty, thirty …, one hundred, none, how many?, down / back, count in ones, twos, threes, fives, tens, many, few, odd, even, every other, how many times?, pattern, pair, ones, tens, exchange, digit, 'teen' numbers, first, second, third … hundredth, last, before, after, next, between, multiple of, sequence, continue, predict

### Monday

Ask each pupil to look at their hand and to count the total number of fingers and thumbs they have (5).

Count together in ones up to 100. Every time you count on, pupils move along the fingers on their hand; when they get to their little finger (or thumb, depending which hand they use), encourage them to shout out the number.

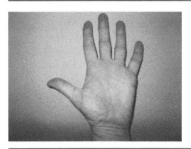

### Tuesday

Repeat Monday's activity, counting up quietly in ones from 0, but shouting the multiples of five. When you reach 100, count backwards in the same way.

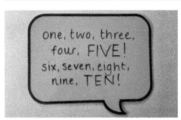

### Wednesday

Ask each pair of pupils to take 20 cubes of one colour and 5 cubes of another colour.

Ask them to lay out the cubes as shown.

Together, count quietly from 0–25; when you reach the different coloured cubes, encourage them to shout out that number. Then count backwards in the same way.

### Thursday

Display a large 100 square.

Together, count quietly from 0–50 in ones, with pupils pointing to their fingers. When you reach a multiple of five, circle the number on the 100 square and encourage pupils to shout out the multiples of five. Then count backwards to 0 in the same way.

Ask pupils what they notice about counting in fives.

### Friday

Remind pupils about Thursday's activity. Ask what was special about counting in fives.

Together, count quietly from 0–100 in ones, with pupils pointing to their fingers. When you reach a multiple of five, circle the number on the 100 square and encourage pupils to shout out the multiples of five. Then count backwards to 0 in same way.

# Week 4: Counting

## Count in tens from any number, forward and backward

**Resources:** cubes, 100 squares

**Vocabulary:** number, numeral, zero, one, two, three …, twenty, thirty …, one hundred, none, how many?, down / back, count in ones, twos, threes, fives, tens, many, few, odd, even, every other, how many times?, pattern, pair, ones, tens, exchange, digit, 'teen' numbers, first, second, third … hundredth, last, before, after, next, between, multiple of, sequence, continue, predict

### Monday

Ask each pupil to look at both of their hands and to count along, wiggling a finger (or thumb) each time you count one more.

Count together in ones, up to 100. Every time you count on, pupils move along the fingers on both of their hands; when they get to a multiple of ten, encourage them to shout out the number.

### Tuesday

Repeat Monday's activity, counting up quietly in ones from 0 and shouting the multiples of ten. When you reach 100, count backwards in the same way.

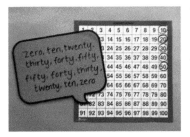

### Wednesday

Ask each pupil to make a tower of 10 cubes.

Ask a pupil to hold a tower of 10. Invite another pupil to join them, and then another until 10 pupils are holding a tower each and there are 100 cubes in total. Count in tens, from 0–100 and then backwards to 0.

### Thursday

Display a large 100 square, or use individual 100 squares.

Together count quietly from 0–50 in ones, with pupils either pointing to the numbers, or using their fingers to count. When you reach a multiple of ten, circle the number on the 100 square and encourage pupils to shout out the multiples of ten. Then count backwards to 0 in same way.

Ask pupils what they notice about counting in tens. (Numbers end in 0.) Finally, count from 0–50 and back again, in tens, using the 100 square to aid fluency.

### Friday

As Thursday, extending to counting to 100. Together count quietly from 0–100 in ones, with pupils using their fingers, or pointing to the numbers on a 100 square. When you reach a multiple of ten, circle the number on the 100 square and encourage pupils to shout out the multiples of ten. Then count backwards to 0 in same way.

Pupils then count in tens from any given number. Finally, count from 0–100 and back again, in tens, using the 100 square to aid fluency.

# Week 5: Counting

**Resources:** Dienes sets, place-value grids, 1–6 dice, large 100 square

**Vocabulary:** number, numeral, zero, one, two, three …, twenty, thirty …, one hundred, none, how many?, down / back, count in ones, twos, threes, fives, tens, many, few, odd, even, every other, how many times?, pattern, pair, ones, tens, exchange, digit, 'teen' numbers, first, second, third … hundredth, last, before, after, next, between, multiple of, sequence, continue, predict

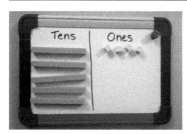

### Monday

Give each pair of pupils a set of Dienes and a place-value grid (or a whiteboard with a grid drawn on).

Roll a dice to establish a start number. Ask pupils to make that number in ones on the place-value grid. Add 10 each time and count aloud up to 100 only. Repeat, rolling the dice for a different start number.

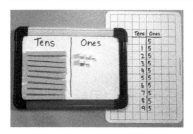

### Tuesday

Give each pair of pupils a set of Dienes and a place-value grid.

Roll a dice to establish a start number. Ask pupils to make that number in ones on the place-value grid. Add 10 each time and count aloud up to 100 only. Then count backwards, removing a rod each time.

If there is time, repeat with different start numbers.

### Wednesday

Working in pairs, pupils repeat Tuesday's activity, rolling their own dice to establish a start number. Give each pair a whiteboard and pen.

Partner 1 in each pair uses Dienes while partner 2 records the numbers.

Ask pupils what they notice about counting in tens. (The ones digit stays the same each time.)

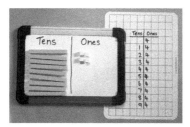

### Thursday

Repeat Wednesday's activity, with partners swapping roles.

Ask how they can check their partner's work. (Check that the ones digit stays the same each time.)

### Friday

Draw a Tens / Ones place-value grid on the board and display a large 100 square.

Choose a one-digit number as the start number. Circle the numbers as you add 10 each time. Ask pupils what patterns they can see when adding tens to any given number. (The ones digit stays the same each time.)

Repeat, choosing a start number between 91 and 99, with pupils counting back 10 each time.

# Week 6: Counting

## Count in steps of 2, 3 and 5 from 0, and in tens from any number, forward and backward

**Resources:** coins, bowls

**Vocabulary:** number, numeral, zero, one, two, three …, twenty, thirty …, one hundred, none, how many?, down / back, count in ones, twos, threes, fives, tens, many, few, odd, even, every other, how many times?, pattern, pair, ones, tens, exchange, digit, 'teen' numbers, first, second, third … hundredth, last, before, after, next, between, multiple of, sequence, continue, predict

### Monday

Give each pupil 10 × 2p coins.

Pupils practise counting them in twos, up to 20p. Then they count backwards to 0.

Pupils then pair up and count up to 40p. Then they count backwards to 0.

### Tuesday

Give each pupil 10 × 5p coins.

Pupils practise counting them in fives, up to 50p. Then they count backwards to 0.

Pupils then pair up and count up to 100p. They then count backwards to 0.

Ask if anyone knows a coin that is equal in value to 100p.

### Wednesday

Exchange two of the 5p coins for a 10p coin, emphasising that two 5p coins have the same value as a 10p coin. Then, give each pupil 10 × 10p coins.

Pupils practise counting them in tens, up to 100p. Highlight that 100p equals one pound. Then they count backwards to 0.

Pupils then pair up; partner 1 in each pair places some coins in a bowl and partner 2 counts them. They check each other's work and then swap over.

### Thursday

Pupils work in pairs; partner 1 in each pair has 9 × 1p coins and partner 2 has 9 × 10p coins.

Partner 1 sets out some coins on the table and partner 2 adds on 10p each time (counting in tens from any given number) and back again. They check each other's work.

### Friday

Repeat Thursday's activity, with pupils swapping roles.

# Week 1: Place value

## Recognise the place value of each digit in a two-digit number (tens and ones)

**Resources:** place-value arrow cards, Dienes sets, bead strings (and similar resources)

> **Vocabulary:** number, numeral, zero, one, two, three …, twenty, thirty …, one hundred, none, count, count up / on / down / back, count in ones, twos … tens, odd, even, every other, how many times?, pattern, pair, ones, tens, exchange, digit, 'teen' numbers, as many as, equal to, greater than / less than, less / more, larger, bigger, less, fewer, smaller, compare, order, size, first, second, third … hundredth, last, before, after, next, between, halfway between, place (value), represents

### Monday

Give each pair of pupils a set of place-value arrow cards and a set of Dienes.

Call out a two-digit number. Partner 1 in each pair shows the number using the place-value arrow cards and partner 2 makes the number using Dienes.

### Tuesday

Repeat Monday's activity, with pupils swapping roles.

### Wednesday

Give each pair of pupils a set of Dienes.

Show a two-digit number using place-value arrow cards. Pairs work together to explore all the different ways in which that number can be partitioned. Highlight that partitioning can be represented in many different ways and that 25 + 7 is just as valuable as 30 + 2. You may choose to emphasise the tens and ones for this activity.

### Thursday

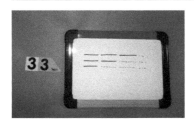

Give each pair of pupils a set of Dienes.

Show a two-digit number using place-value arrow cards. Pairs work together to find several different ways in which that number can be partitioned. These should be recorded in any way pupils choose – this may be a bead string, number line, representations, Dienes, etc.

### Friday

Give each pair of pupils a whiteboard and pen.

Show a two-digit number using place-value arrow cards. Pairs work together to find all the different ways in which that number can be partitioned. Pupils should record the different ways on a whiteboard.

# Week 2: Place value

## Recognise the place value of each digit in a two-digit number (tens and ones)

**Resources:** place-value arrow cards, 1–6 dice

**Vocabulary:** number, numeral, zero, one, two, three …, twenty, thirty …, one hundred, none, count, count up / on / down / back, count in ones, twos … tens, odd, even, every other, how many times?, pattern, pair, ones, tens, exchange, digit, 'teen' numbers, as many as, equal to, greater than / less than, less / more, larger, bigger, less, fewer, smaller, compare, order, size, first, second, third … hundredth, last, before, after, next, between, halfway between, place (value), represents

### Monday

Give each pair of pupils a set of place-value arrow cards.

Say a two-digit number. Ask pupils to show you the value, using place-value arrow cards. Repeat with different numbers.

### Tuesday

Give each pair of pupils a set of place-value arrow cards.

Roll a dice. Pupils use the number to make two different two-digit numbers using place-value arrow cards. For example, if you roll a 6, pupils make one two-digit number that uses a 6 in the tens column and another two-digit number that uses a 6 in the ones column. Repeat the activity.

### Wednesday

Give each pupil a set of place-value arrow cards.

Roll two different coloured dice. Explain that one colour represents tens and the other colour represents ones. Ask pupils to use the place-value arrow cards to show the number that has been rolled on the dice. Repeat the activity.

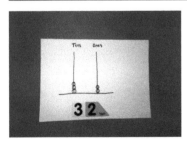

### Thursday

Give each pair of pupils a whiteboard or sheet of paper and a pen.

Model drawing a tens and ones abacus on the whiteboard.

Show a number using the place-value arrow cards. Ask pairs to draw that number on an abacus on their whiteboards and to show you. Repeat the activity.

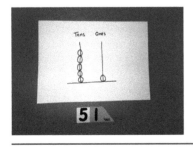

### Friday

Give each pupil a set of place-value arrow cards.

Represent a number using an abacus.

Ask pupils to show that number using the place-value arrow cards. Repeat the activity.

# Week 3: Place value

## Read and write numbers to at least 100 in numerals and words

**Resources:** Dienes sets, 1–100 number word cards, place-value arrow cards

**Vocabulary:** number, numeral, zero, one, two, three …, twenty, thirty …, one hundred, none, count, count up / on / down / back, count in ones, twos … tens, odd, even, every other, how many times?, pattern, pair, ones, tens, exchange, digit, 'teen' numbers, as many as, equal to, greater than / less than, less / more, larger, bigger, less, fewer, smaller, compare, order, size, first, second, third … hundredth, last, before, after, next, between, halfway between, place (value), represents

---

### Monday

Give each pair of pupils a set of Dienes and a handful of cards showing single-digit numerals and multiples of ten as words.

Pupils take two cards, read the number together and show it using Dienes.

---

### Tuesday

Give each pair of pupils a set of Dienes, a handful of number word cards showing ones numbers and multiples of ten, and a whiteboard and pen.

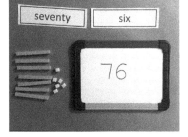

Pupils take two cards and read the number together. Partner 1 in each pair shows that number using Dienes and partner 2 writes the number on a whiteboard. They check each other's work.

---

### Wednesday

Repeat Tuesday's activity, with pupils swapping roles. They check each other's work.

---

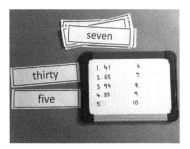

### Thursday

Give each pupil a whiteboard and pen.

Choose two number word cards and show them to the class. Pupils write the number in **numerals**. Give ample thinking time for pupils to read the words and write the number. Repeat several times. Allow time for pupils to mark their own work.

---

### Friday

Give each pupil a whiteboard and pen.

Show a number using place-value arrow cards. Pupils must write the number in **words**. Give ample thinking time for pupils to read the numerals and to write the words. Repeat several times. Allow time for pupils to mark their own work.

Pupils could create a class word bank, add to a working wall or create a poster.

---

# Week 4: Place value

## Compare and order numbers from 0 up to 100; use = sign

**Resources:** lolly sticks, cubes or Dienes sets

**Vocabulary:** number, numeral, zero, one, two, three …, twenty, thirty …, one hundred, none, count, count up / on / down / back, count in ones, twos … ten, odd, even, every other, how many times?, pattern, pair, ones, tens, exchange, digit, 'teen' numbers, as many as, equal to, greater than / less than, less / more, larger, bigger, less, fewer, smaller, compare, order, size, first, second, third … hundredth, last, before, after, next, between, halfway between, place (value), represents

### Monday

Give each pupil two lolly sticks.

Ask pupils to make the = sign with the sticks. Demonstrate that an equal number of cubes must lie between the sticks so that the sticks are the same distance from each other. Explain that we use the equals symbol to show that one value is equivalent in value to another. Repeat with different quantities of cubes.

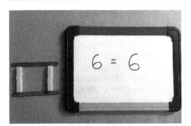

### Tuesday

Demonstrate writing a sentence using the = sign.

Ask pupils: if 6 is on one side of the = sign, what value will be on the other? Demonstrate exploring all the possibilities: $6 = 0 + 6$, $6 = 1 + 5$, $6 = 2 + 4$, $6 = 3 + 3$, $6 = 4 + 2$, $6 = 5 + 1$, $6 = 6 + 0$. Repeat for other single-digit numbers.

### Wednesday

Repeat Tuesday's activity with increasingly large numbers. Begin with a teens number and then move to numbers between 20 and 50. Encourage pupils to 'make' the numbers to check.

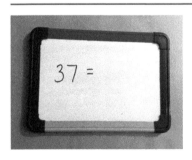

### Thursday

Give each pair of pupils a whiteboard and pen.

Write and call out a target number between 11 and 99. Pairs explore as many different ways as they can to make a correct number sentence. (Encourage them to write some number sentences using the $\square = x$ format.) These should be recorded on a whiteboard. Pupils can use cubes or Dienes or they may be confident without this support.

### Friday

Give each pair of pupils a whiteboard and pen.

Write and call out a target number between 51 and 99. Pairs explore as many different ways as they can to make a correct number sentence. (Encourage them to write some number sentences using the $\square = x$ format.) These should be recorded on a whiteboard.

# Week 5: Place value

## Compare and order numbers from 0 up to 100; use < sign

**Resources:** lolly sticks, objects, cubes

**Vocabulary:** number, numeral, zero, one, two, three …, twenty, thirty …, one hundred, none, count, count up / on / down / back, count in ones, twos … tens, odd, even, every other, how many times?, pattern, pair, ones, tens, exchange, digit, 'teen' numbers, as many as, equal to, greater than / less than, less / more, larger, bigger, less, fewer, smaller, compare, order, size, first, second, third … hundredth, last, before, after, next, between, halfway between, place (value), represents

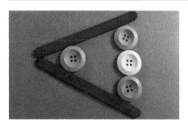

### Monday

Give each pupil two lolly sticks.

Ask pupils to make the < sign with the sticks. Demonstrate that only one object will fit in the vertex, but that several objects will fit in the open end. Explain that we use this symbol to represent 'is less than': 1 is less than 3, or 1 < 3.

### Tuesday

Give each pair of pupils two lolly sticks.

Remind pupils that only a small number of objects will fit in the vertex of the < sign, but that several objects will fit in the open end. This symbol represents 'is less than'.

Ask pairs to make the < sign using the sticks. Partner 1 in each pair places a small number of objects (less than 10) at one side of the < sign. Partner 2 places an appropriate number of objects at the other side.

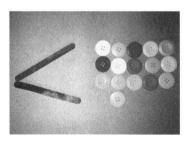

### Wednesday

Repeat Tuesday's activity, with pupils exploring how many different combinations there may be when quantities of objects are placed at different sides of the < sign.

Probe understanding by asking pupils to explain how many objects they think there might be at the vertex if there are 17 objects at the open end. Repeat with increasingly larger numbers.

### Thursday

Give each pair of pupils two lolly sticks and a whiteboard and pen. They will make the < sign using the sticks and put it on a whiteboard to make the connection between the concrete and the written form.

Partner 1 in each pair forms the < sign and then writes a number at one side. Partner 2 then writes an appropriate number at the other side to make the statement correct. Pupils practise saying the sentence aloud.

### Friday

Repeat Thursday's activity, with pupils swapping roles.

# Week 6: Place value

## Compare and order numbers from 0 up to 100; use > sign

**Resources:** lolly sticks, objects, cubes

**Vocabulary:** number, numeral, zero, one, two, three …, twenty, thirty …, one hundred, none, count, count up / on / down / back, count in ones, twos … tens, odd, even, every other, how many times?, pattern, pair, ones, tens, exchange, digit, 'teen' numbers, as many as, equal to, greater than / less than, less / more, larger, bigger, less, fewer, smaller, compare, order, size, first, second, third … hundredth, last, before, after, next, between, halfway between, place (value), represents

### Monday

Give each pupil two lolly sticks.

Ask pupils to make the > sign using the sticks. Demonstrate that only a few objects will fit in the vertex, but that several more objects will fit in the open end. Explain that we use this symbol to represent 'is greater than': 4 is greater than 2, or 4 > 2.

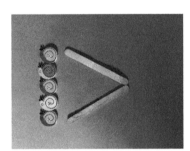

### Tuesday

Give each pair of pupils two lolly sticks.

Remind pupils that only a small number of objects will fit in the vertex of the > sign, but that several more objects will fit in the open end. This symbol represents 'is greater than'.

Ask pairs to make the > sign using the sticks. Partner 1 in each pair places a number of objects (less than 10) at one side of the > sign. Partner 2 places an appropriate number of objects at the other side to make the statement correct.

### Wednesday

Repeat Tuesday's activity, with pupils exploring how many different combinations there may be when the objects are placed in different areas with the > sign.

Probe understanding by asking pupils to explain how many objects they think there might be at the open end if there are 2 objects at the vertex.

### Thursday

Give each pair of pupils two lolly sticks and a whiteboard and pen. They will make the > sign using the sticks and put it on a whiteboard to make the connection between the concrete and the written form.

Partner 1 in each pair forms the > sign and then writes a number at one side. Partner 2 then writes an appropriate number at the other side to make the statement correct. Pupils should practise saying the sentence aloud.

### Friday

Repeat Thursday's activity, with pupils swapping roles.

# Week 1: Representing numbers

## Identify and represent numbers 0–100 using different representations

**Resources:** objects, ten frames, matchsticks/straws, 1–6 dice

> **Vocabulary:** number, numeral, zero, one, two, three ..., twenty, thirty ..., one hundred, none, count, count up / on / down / back, count in ones, twos ... tens, odd, even, every other, how many times?, pattern, pair, ones, tens, exchange, digit, `teens' number, as many as, equal to, greater than / less than, less / more, larger, bigger, less, fewer, smaller, compare, order, size, first, second, third ... hundredth, last, before, after, next, between, halfway between, place (value), represents

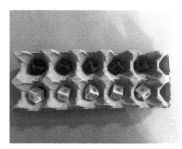

### Monday

Give each pair of pupils a set of 10 objects and a ten frame (or a printed frame, an egg box or an ice-cube tray).

Explain how the ten frame works: the top row is filled first, and when it is full there will be 5 objects. Then the bottom row is filled, and when it is full there will be 5 objects. When both rows are filled, there will be 5 + 5 objects, that is 10 objects in total.

Ask pupils to fill the ten frame correctly to explore this through counting.

### Tuesday

Give each pupil a ten frame, a set of 10 objects, 5 matchsticks/straws and a dice.

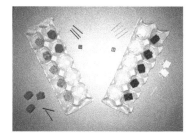

Playing in pairs, pupils take turns to roll a dice and fill their ten frame accordingly (e.g. roll a 5 and put 5 objects in the frame). Once their frame is full, they exchange the 10 objects for a matchstick and start again. If they roll a 6 but have only two spaces in the ten frame, they would need to exchange 10 objects for a matchstick and start the empty frame with the remaining 4 objects. The first person to collect 5 matchsticks (representing 50) is the winner.

### Wednesday

Repeat Tuesday's activity to practise the exchange. Pupils could work towards a different target number (e.g. 70).

### Thursday

Give each pair of pupils a set of 10 objects and a ten frame.

Explain that today each space in the ten frame is worth 10 rather than 1.

Call out a target multiple of ten (e.g. 70). Pupils represent that number using the ten frame and the objects.

### Friday

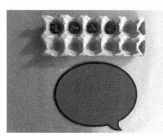

Give each pair of pupils a set of 10 objects and a ten frame.

Explain that today each space in the ten frame is worth 10 rather than 1.

To practise their knowledge of related number bond facts, say a multiple of ten (e.g. 60) and ask them to show you the corresponding number that would total 100.

# Week 2: Representing numbers

## Identify and represent numbers 0–100 using different representations

**Resources:** matchsticks, bowls, 100 squares

**Vocabulary:** number, numeral, zero, one, two, three …, twenty, thirty …, one hundred, none, how many?, count, count up / on / down / back, count in ones, twos … tens, odd, even, every other, how many times?, pattern, pair, ones, tens, exchange, digit, 'teen' numbers, the same number as, as many as, equal to, greater than / less than, less / more, larger, bigger, less, fewer, smaller, compare, order, size, first, second, third … tenth, eleventh … twentieth … hundredth, last, before, after, next, between, halfway between, place (value), represents

### Monday

Give each pupil a handful of matchsticks.

Demonstrate using a tally to represent numbers with the matchsticks.

Together, count out 15 matchsticks, representing fives as a tally. When you have positioned the matchsticks, count together in fives to 15. Repeat for 20, 25 and 30.

### Tuesday

Give each pair of pupils a bowl full of matchsticks and a 100 square.

Partner 1 in each pair circles a multiple of five between 1 and 50 on the 100 square. Partner 2 then represents that number using a tally with matchsticks. Partner 1 counts in fives to check the tally is correct. Repeat the activity.

### Wednesday

Repeat Tuesday's activity, with pupils swapping roles.

### Thursday

Give each pair of pupils a 100 square and a whiteboard and pen.

Model drawing a tally using lines on a whiteboard.

Partner 1 in each pair circles a multiple of five between 50 and 100 on the 100 square. Partner 2 then represents that number using a tally. Partner 1 counts in fives to check the tally is correct. Repeat the activity.

### Friday

Repeat Thursday's activity, with pupils swapping roles.

# Week 3: Representing numbers

## Identify and represent numbers 0–100 using number lines

**Resources:** blank number lines (or strips of paper), 100 squares, counters

> **Vocabulary:** number, numeral, zero, one, two, three …, twenty, thirty …, one hundred, none, count, count up / on / down / back, count in ones, twos … tens, odd, even, every other, how many times?, pattern, pair, ones, tens, exchange, digit, 'teen' numbers, as many as, equal to, greater than / less than, less / more, larger, bigger, less, fewer, smaller, compare, order, size, first, second, third … hundredth, last, before, after, next, between, halfway between, place (value), represents

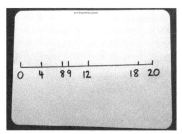

### Monday

Give each pupil either a blank strip of paper as a number line, or use a whiteboard. Ask pupils to mark 0 and 20 on the number line.

Call out a number between 0 and 20 and ask pupils to draw a line on the number line to represent the number. Give between five and eight numbers in total to place on the number line.

Ask pupils to compare (and reason) as to their positioning of numbers on the number line.

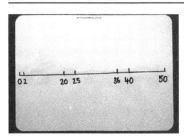

### Tuesday

Give each pupil either a blank strip of paper as a number line, or use a whiteboard. Ask pupils to mark 0 and 50 on the number line.

Call out a number between 0 and 50 and ask pupils to draw a line on the number line to represent the number. Repeat.

Ask pupils to compare (and reason) as to their positioning of numbers on the number line.

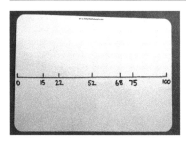

### Wednesday

Give each pupil either a blank strip of paper as a number line, or use a whiteboard. Ask pupils to mark 0 and 100 on the number line.

Call out a number between 0 and 100 and ask pupils to draw a line on the number line to represent the number. Repeat.

Ask pupils to compare (and reason) as to their positioning of numbers on the number line.

### Thursday

Give each pair of pupils a blank 1–100 number line, a 100 square and some counters.

Partner 1 in each pair uses counters to cover five numbers on the 100 square. Together they discuss then mark those numbers on the number line.

### Friday

Repeat Thursday's activity, with partner 2 covering five numbers.

# Week 4: Representing numbers

## Estimate numbers 0–50

**Resources:** objects, number lines, 100 squares

**Vocabulary:** number, numeral, zero, one, two, three …, twenty, thirty …, one hundred, none, count, count up / on / down / back, count in ones, twos … tens, odd, even, every other, how many times?, pattern, pair, ones, tens, exchange, digit, 'teen' numbers, as many as, equal to, greater than / less than, less / more, larger, bigger, less, fewer, smaller, compare, order, size, first, second, third … hundredth, last, before, after, next, between, halfway between, place (value), represents

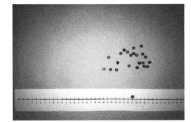

### Monday

Give each pair of pupils a number of small objects and a number line (minimum 0–50).

Partner 1 in each pair places a handful of objects on the table and counts 5 seconds. Partner 2 circles on the number line their estimate of the number of objects shown in total. They check the answer together by counting and grouping in tens.

### Tuesday

Repeat Monday's activity, with pupils swapping roles.

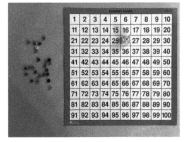

### Wednesday

Give each pair of pupils a number of small objects and a 100 square.

Partner 1 in each pair places a handful of objects on the table, places a cube on their estimate of the quantity on the 100 square and then counts 5 seconds for partner 2 to say whether they think the answer is greater than, less than or equal to the number indicated. They check the answer together by counting and grouping in tens.

### Thursday

Repeat Wednesday's activity, with pupils swapping roles.

### Friday

Give each pair of pupils a number of small objects and a whiteboard and pen.

Each pupil takes a quantity of objects and places them on a whiteboard in two separate piles. Together they estimate the number of objects in each pile, and then write <, > or = to make the statement correct. They check the answer together by counting and grouping in tens.

# Week 5: Representing numbers

## Estimate numbers 0–100

**Resources:** containers, objects, 100 squares / number lines

**Vocabulary:** number, numeral, zero, one, two, three …, twenty, thirty …, one hundred, none, count, count up / on / down / back, count in ones, twos … tens, odd, even, every other, how many times?, pattern, pair, ones, tens, exchange, digit, 'teen' numbers, as many as, equal to, greater than / less than, less / more, larger, bigger, less, fewer, smaller, compare, order, size, first, second, third … hundredth, last, before, after, next, between, halfway between, place (value), represents

### Monday

Give each pair of pupils a cup. Make a selection of different objects available.

Pupils work in pairs to estimate how many of the different objects will fill their cup (e.g. how many cubes will fill the cup to the top or how many mini-beast toys will fill the cup) and then count to check. To help with accuracy, you could show how much of the cup is filled with 5 objects and then use that as a benchmark for estimating.

### Tuesday

Give each pair of pupils a container. Make a selection of different objects available.

Together they estimate how many of the different objects will fill the container (e.g. how many beads will fill the container to the top or how many toy cars will fill the container) and then count to check.

### Wednesday

Give each pair of pupils a number of small objects (between 1 and 100) and a 100 square / number line.

Partner 1 in each pair estimates how many objects there are and partner 2 records the estimate on a 100 square / number line or as a tally. They check their answer together by counting.

### Thursday

Repeat Wednesday's activity, with pupils swapping roles.

### Friday

Give each pair of pupils a 100 square and a large quantity of small objects.

Partner 1 in each pair circles a number on the 100 square. Partner 2 has 10 seconds to use the objects to represent their estimate of that quantity. Partner 1 then checks the accuracy of the estimate by counting the objects. Pupils swap roles.

# Week 6: Representing numbers

## Compare and order numbers from 0 up to 100; use <, > or = signs

**Resources:** cups, cubes, lolly sticks

**Vocabulary:** number, numeral, zero, one, two, three ..., twenty, thirty ..., one hundred, none, count, count up / on / down / back, count in ones, twos ... tens, odd, even, every other, how many times?, pattern, pair, ones, tens, exchange, digit, 'teen' numbers, as many as, equal to, greater than / less than, less / more, larger, bigger, less, fewer, smaller, compare, order, size, first, second, third ... hundredth, last, before, after, next, between, halfway between, place (value), represents

### Monday

Give each pupil a cup of cubes (between 1 and 100) and two lolly sticks.

Working in pairs, pupils estimate the number of cubes and whether partner 1's cube total is less than, equal to or more than partner 2's. They use the lolly sticks to make the statement correct using the sign <, > or =. Pupils check their statement together by counting the cubes.

### Tuesday

Give each pair of pupils a cup of cubes (between 1 and 100), two lolly sticks and a whiteboard and pen.

Pupils estimate the number of cubes and use the lolly sticks to create the < sign. They then write a number on the other side on a whiteboard to make the statement correct. Pupils check their statement together by counting the cubes.

### Wednesday

Give each pair of pupils a cup of cubes (between 1 and 100), two lolly sticks and a whiteboard and pen.

Pupils estimate the number of cubes and use the lolly sticks to create the > sign. They then write a number on the other side on a whiteboard to make the statement correct. Pupils check their statement together by counting the cubes.

### Thursday

Give each pair of pupils a cup of cubes (between 1 and 100), two lolly sticks and a whiteboard and pen.

Pupils estimate the number of cubes and use the lolly sticks to create the = sign. They then write a number on the other side on a whiteboard to make the statement correct. Pupils check their statement together by counting the cubes.

### Friday

Give each pair of pupils a cup of cubes (between 1 and 100), two lolly sticks and a whiteboard and pen.

Pupils estimate the number of cubes and use the lolly sticks to create the <, > or = sign. They then write a tally to represent a number on the other side to make the statement correct. Pupils check their statement together by counting the cubes.

# Week 1: Addition and subtraction

## Recall and use addition and subtraction facts to 20 fluently

**Resources:** dominoes

**Vocabulary:** +, add, addition, more, more than, plus, make, sum, total, altogether, how many more to make …?, how many more is … than …?, how much more is …?, −, subtract, take / take away, minus, less, one less, two less … ten less, how many fewer is … than …?, how much less is … than …?, what is the difference between … and …?, =, equals / equal to, is the same as, sign, symbol, tens, ones, place value

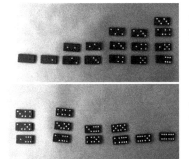

### Monday

Give a set of dominoes to a group of four pupils.

Ask the group to find the spot totals and place them in order from 0–12. One pair should find totals 0–6 and the other pair should find totals 7–12.

### Tuesday

Give a set of dominoes to a group of four pupils.

Repeat Monday's activity, swapping roles so that both pairs find all combinations to 12.

### Wednesday

Pupils could work in twos or fours for this activity.

Pupils take a set of dominoes and start with all the dominoes face up. One player chooses a domino as the 'target total' domino and the other pupils find another domino that has the same spot total. Pupils then take turns to say the next 'target total' for their group to find.

### Thursday

Give each pair of pupils a handful of dominoes and a whiteboard and pen.

Pupils start with all the dominoes face down. One pupil turns over two dominoes. The pupils then work together to write a subtraction number sentence using those spot totals.

Repeat, with pupils taking turns to turn over two dominoes.

### Friday

Give each pair of pupils a handful of dominoes (take out the dominoes with spot totals of 12 and 11) and a whiteboard and pen.

Pupils start with all the dominoes face down. One pupil turns over two dominoes. The pupils then add the total of the spots, and deduct the total from 20, writing this as a number sentence.

Repeat, with pupils taking turns to turn over two dominoes.

# Week 2: Addition and subtraction

## Recall and use addition and subtraction facts to 20 fluently, and derive and use related facts up to 100

**Resources:** pegs on coat hangers / bead strings (or similar resource with movable objects)

**Vocabulary:** +, add, addition, more, more than, plus, make, sum, total, altogether, how many more to make …?, how many more is … than …?, how much more is …?, −, subtract, take / take away, minus, less, one less, two less … ten less, how many fewer is … than …?, how much less is … than …?, what is the difference between … and …?, =, equals / equal to, is the same as, sign, symbol, tens, ones, place value

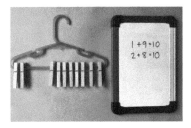

### Monday

Give each pair of pupils 10 pegs on a coat hanger (or similar) and a whiteboard and pen.

Challenge them to find all of the ways in which 10 can be made from two numbers. Encourage sliding 1 peg across to show 10 = 1 + 9, then 10 = 2 + 8, then 10 = 3 + 7, etc. Discuss 10 + 0 and 0 + 10. Pupils record these, if wished, and display throughout the week. Encourage a systematic method.

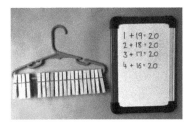

### Tuesday

Give each pair of pupils 20 pegs on a coat hanger and a whiteboard and pen.

Challenge them to find all of the ways in which 20 can be made from two numbers. Remind them of Monday's activity and encourage sliding 1 peg across to show 20 = 1 + 19, then 20 = 2 + 18, then 20 = 3 + 17, etc. Pupils record these and display throughout the week. Encourage a systematic method.

### Wednesday

Give each pair of pupils 20 pegs on a coat hanger.

Play 'I say, you say' where you call out a number between 0 and 20 (e.g. 12). Pupils slide that number of pegs along, count the remaining pegs and chant (e.g. 'You say 12, we say 8! 12 + 8 = 20'). Repeat.

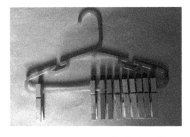

### Thursday

Give each pair of pupils 10 pegs on a coat hanger and explain that, this time, each peg represents 10. Count in tens, from 0–100, sliding pegs across each time.

Challenge pupils to find and record all of the addition bonds to 100. Start by sliding one peg across and showing that 10 + 90 = 100, then 20 + 80 = 100, etc. Encourage a systematic method. Once you have finished, pupils can record these.

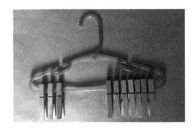

### Friday

Give each pair of pupils 10 pegs on a coat hanger and explain that, as on Thursday, each peg represents 10. Count back from 100–0, in tens, sliding pegs across each time.

Challenge pupils to find and record all of the subtraction bonds to 100. Start by sliding one peg across and showing that 100 − 10 = 90, then 100 − 20 = 80, etc. Encourage a systematic method. Once you have finished, pupils can record these.

# Week 3: Addition and subtraction

## Add and subtract a two-digit number and ones

**Resources:** Cubes, 1–6 dice, number line (optional)

> **Vocabulary:** +, add, addition, more, more than, plus, make, sum, total, altogether, how many more to make …?, how many more is … than …?, how much more is …?, –, subtract, take / take away, minus, less, one less, two less … ten less, how many fewer is … than …?, how much less is … than …?, what is the difference between … and …?, =, equals / equal to, is the same as, sign, symbol, tens, ones, place value

### Monday

Give each pupil 30 cubes, a dice and a whiteboard and pen.

Ask pupils to write the numbers 21–30 on their whiteboard, using a number line for support if needed.

Starting at 21, pupils roll the dice and add the number shown by counting on from 21. (Encourage counting the spots by touching them.) Next, roll the dice again and add that spot total to 22, by counting on and touching the spots. Pupils should record these as number sentences using + and = signs on their whiteboards. Model how to use the cubes to check their answers.

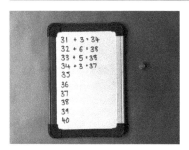

### Tuesday

Repeat Monday's activity, with pupils writing the numbers 31–40 on their whiteboard.

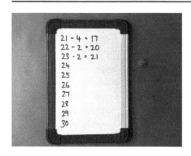

### Wednesday

Give each pupil 30 cubes, a dice and a whiteboard and pen.

Ask pupils to write the numbers 21–30 on their whiteboard.

Starting at 21, pupils roll the dice and subtract the number shown from 21 by counting back from 21. (Encourage them to count back touching the spots.) Repeat with the start number as 22. Model how to use the cubes to check their answers.

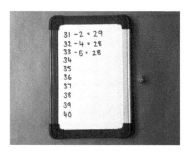

### Thursday

Repeat Wednesday's activity, with pupils writing the numbers 31–40 on their whiteboard.

### Friday

Repeat Thursday's activity, with pupils writing the numbers 41–50 on their whiteboard (or choose the start numbers that are best suited to your class).

# Week 4: Addition and subtraction

## Add and subtract a two-digit number and tens

**Resources:** 0–100 number line / metre stick / tape measure, large 100 square

> **Vocabulary:** +, add, addition, more, more than, plus, make, sum, total, altogether, how many more to make …?, how many more is … than …?, how much more is …?, −, subtract, take / take away, minus, less, one less, two less … ten less, how many fewer is … than …?, how much less is … than …?, what is the difference between … and …?, =, equals / equal to, is the same as, sign, symbol, tens, ones, place value

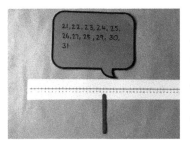

### Monday

Give each pair of pupils a 0–100 number line (or a metre stick or tape measure). It would be useful to display a 100 square.

Call out a start number between 11 and 39 and ask pupils to point to that number on their resource. One pupil keeps their finger on that start number. The other pupil must count on 10. What number do they land on?

Repeat for different start numbers to practise adding 10, by counting on 10 ones.

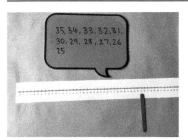

### Tuesday

Give each pair of pupils a 0–100 number line. It would be useful to display a 100 square.

Call out a start number between 21 and 49 and ask pupils to point to that number. One pupil keeps their finger on that start number. The other pupil must count back 10. What number do they land on?

Repeat for different start numbers to practise subtracting 10 by counting back 10 ones.

### Wednesday

Repeat Monday's activity with a start number between 51 and 89.

### Thursday

Repeat Tuesday's activity with a start number between 31 and 49.

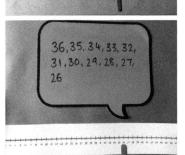

### Friday

Repeat Thursday's activity, with a start number between 51 and 99.

# Week 5: Addition and subtraction

## Add and subtract two two-digit numbers

**Resources:** base 10 / Dienes (or similar, e.g. lolly sticks and cubes), large 100 square, place-value arrow cards

> **Vocabulary:** +, add, addition, more, more than, plus, make, sum, total, altogether, how many more to make …?, how many more is … than …?, how much more is …?, −, subtract, take / take away, minus, less, one less, two less … ten less, how many fewer is … than …?, how much less is … than …?, what is the difference between … and …?, =, equals / equal to, is the same as, sign, symbol, tens, ones, place value

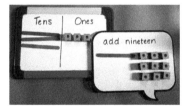

### Monday

Throughout the week each pair of pupils will need a set of concrete objects to represent tens and ones and a whiteboard and pen.

On a whiteboard draw a line down the middle and write 'tens' on the left and 'ones' on the right. Point to a number on a 100 square and ask pupils to represent that number using the concrete objects. Repeat to revise place value.

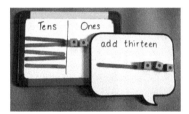

### Tuesday

Give each pair of pupils their set of tens and ones objects and a whiteboard and pen.

Call out (or show using place-value arrow cards) a two-digit number and ask pupils to make that number using their resource. Ask pupils to add a 'teens number' to that number (do not bridge across tens). Model adding a ten and then the ones. Ask pupils to find the total of the combined tens and ones.

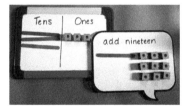

### Wednesday

Repeat Tuesday's activity, with pupils bridging across tens, and exchanging 12 cubes for a ten and 2 ones.

Use the language of exchange. Allow time for pupils to practise.

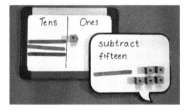

### Thursday

Give each pair of pupils their set of tens and ones objects and a whiteboard and pen.

Call out (or show using place-value arrow cards) a two-digit number and ask pupils to make that number using their resource. Ask pupils to subtract a 'teens number' from that number. Avoid bridging across 10. Model subtracting a ten and then the ones. Ask pupils to find the total of the remaining tens and ones.

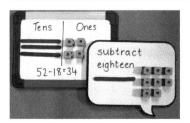

### Friday

Repeat Thursday's activity, with pupils recording answers using a number sentence on a whiteboard. When confident, pupils can bridge across 10.

## Add three one-digit numbers

**Resources:** ten frames (or a printed frame, an egg box, an ice-cube tray), 1–6 dice, cubes

**Vocabulary:** +, add, addition, more, more than, plus, make, sum, total, altogether, how many more to make …?, how many more is … than …?, how much more is …?, –, subtract, take / take away, minus, less, one less, two less … ten less, how many fewer is … than …?, how much less is … than …?, what is the difference between … and …?, =, equals / equal to, is the same as, sign, symbol, tens, ones, place value

### Monday

Throughout the week each group of three pupils will need two ten frames of any form, three dice and some cubes.

Each pupil rolls their dice and places that number of cubes in the ten frame (as they learnt in the 'Representing numbers' section) to find the total of three one-digit numbers.

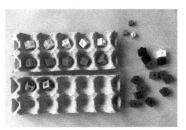

### Tuesday

Each pupil in each group of three needs a set of 6 cubes, with each pupil choosing a different colour. Each group needs two ten frames and three dice.

As in Monday's activity, each pupil rolls their dice and places their cubes in the ten frame to find the total of three one-digit numbers. By using different colours, pupils may see that numbers are partitioned to suit the addition.

### Wednesday

Each pupil in each group of three needs a set of 6 cubes, with each pupil choosing a different colour.

Each pupil rolls their dice and the group discusses if they can manipulate the numbers to make number bonds within 10. For example, if they roll a 4, a 2 and a 5, pupils might say that 5 + 2 + 3 makes 10, with one left over (seeing the 4 as 3 + 1).

### Thursday

Each pupil in each group of three needs a set of 6 cubes, with each pupil choosing a different colour. Give each group a whiteboard and pen.

Set a target total (between 5 and 18) and ask each group to make that number using their cubes. Pupils record as many ways as possible to find the target total. Pupils should explore the commutativity of addition, using the vocabulary of commutativity.

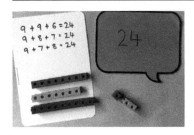

### Friday

Pupils work in groups of three and have a tower of 9 cubes each. As in Wednesday's activity, pupils explore and make number bonds to develop fluency. For example, with 9 + 9 + 6, pupils could take two cubes from the tower of 6 and add to the two 9 towers to make 10 + 10 + 4. Pupils record their explorations.

# Week 1: Multiplication and division

## Recall and use multiplication and division facts for the 2 multiplication table

**Resources:** objects, Cuisenaire rods / cubes

**Vocabulary:** ×, lots of, groups of, times, multiply, multiplied by, multiple of, once, twice, three times … twelve times, *n* times as (big, long, wide, etc.), repeated addition, array, row, column, double, ÷, halve, share, share equally, one each, two each, three each …, group in pairs, threes … tens, equal groups of, divide, divided by, divided into, left, left over

---

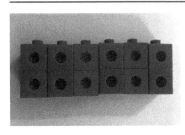

### Monday

Invite pupils to use objects that interest them.

Write a 2 multiplication table fact on the board and ask pupils to prove it in as many different ways as they can. Take photographs of their work to display throughout the week.

---

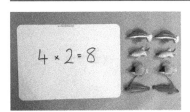

### Tuesday

Give each pair of pupils Cuisenaire rods or some cubes and a whiteboard (or squared paper).

Ask pupils to put the cubes in twos to create a 6 × 2 array. Ask them to describe to each other what they can see in the arrangement. Ask pupils to rotate the array a quarter turn to see if it makes any difference to what they see. Ask if they can see 6 × 2 and 2 × 6. Encourage discussion using correct vocabulary.

---

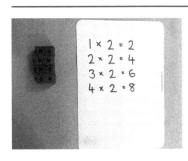

### Wednesday

Give each pair of pupils either red Cuisenaire twos rods or some cubes linked in twos and a whiteboard (or paper) and pen.

Ask pupils to show you an arrangement of 4 groups of 2. Ask if 4 groups of 2 is the same as 2 groups of 4. Explore further facts (e.g. 7 groups of 2, 8 groups of 2). This will support their understanding of commutativity. Pupils should write number sentences together.

---

### Thursday

Give each pair of pupils either red Cuisenaire twos rods or cubes linked in twos and a whiteboard (or paper) and pens.

Starting at 1 × 2, ask partner 1 in each pair to create the array while partner 2 records the formal multiplication table facts. Each time, ask partner 1 to rotate the array and give the related fact. This could be recorded.

---

### Friday

Repeat Thursday's activity, with pupils swapping roles.

---

**Recall and use multiplication and division facts for the 2 multiplication table, including recognising odd and even numbers**

**Resources:** Cuisenaire rods / cubes

**Vocabulary:** × lots of, groups of, times, multiply, multiplied by, multiple of, once, twice, three times … twelve times, *n* times as (big, long, wide, etc.), repeated addition, array, row, column, double, ÷, halve, share, share equally, one each, two each, three each …, group in pairs, threes … tens, equal groups of, divide, divided by, divided into, left, left over

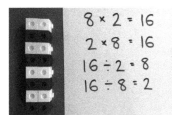

### Monday

Give each pair of pupils Cuisenaire rods or some cubes and a whiteboard (or paper) and pen.

Ask pupils to create a 2 × 8 array using cubes and to write both facts (2 × 8 = 16 and 8 × 2 = 16). Now ask pupils how many rows of 2 there are in the array of 16. Model writing 16 ÷ 2 = 8. Now rotate the array and ask pupils how many rows of 8 there are in 16. Model writing 16 ÷ 8 = 2.
Discuss 16 = 2 × 8, 16 = 8 × 2, 8 = 16 ÷ 2, 2 = 16 ÷ 8.

### Tuesday

Give each pair of pupils Cuisenaire rods or some cubes and a whiteboard (or paper) and pen.

Call out a multiple of two as a multiplication table fact and ask partner 1 in each pair to make an array that demonstrates this. Partner 2 records both multiplication facts relating to the array. Then ask pupils to write the related division facts together.

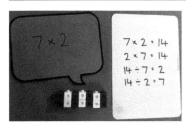

### Wednesday

Give each pair of pupils 24 cubes linked in twos (i.e. 12 pairs).

Call out a multiple of two between 2 and 24 and ask pupils to arrange the cubes in pairs as an array. Explain that they have divided *x* cubes into twos and they have now got *x* groups of 2 cubes. Repeat for other multiples of two.

### Thursday

Give each pair of pupils 25 cubes (enough to make 12 sets of 2 cube pairs, plus 1 single cube) and a whiteboard and pen.

Explain that they are going to make all the numbers from 1–24, one at a time, using pairs of cubes and ones. Explain that even numbers are multiples of two but odd numbers are not. Pupils could take it in turns to make the numbers or to record. It may be helpful to have the single cube a different colour to the other cubes to aid understanding.

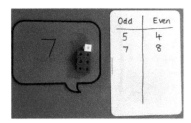

### Friday

Give each pair of pupils 25 cubes (enough to make 12 sets of 2 cubes linked together plus 1 single cube) and a whiteboard and pen.

Pupils draw a line down the middle of their whiteboard and write 'odd' and 'even' at the top. Call out a number between 1 and 24. Ask pupils to make the number and then record whether it is even or odd. Repeat for other numbers.

# Week 3: Multiplication and division

**Vocabulary:** × lots of, groups of, times, multiply, multiplied by, multiple of, once, twice, three times … twelve times, n times as (big, long, wide, etc.), repeated addition, array, row, column, double, ÷, halve, share, share equally, one each, two each, three each …, group in pairs, threes … tens, equal groups of, divide, divided by, divided into, left, left over

### Monday

Start by counting in fives from 0–60 and back again, using either a counting stick or a 100 square to support.

Ask each pair of pupils to make 3 towers of 5 cubes.

Write '3 groups of 5 equals 15' on the board and ask pupils to prove it in as many different ways as they can. Take photographs to display throughout the week.

### Tuesday

Start by counting in fives from 0–60 and back again, using either a counting stick or a 100 square to support.

Give each pair of pupils 12 towers of 5 cubes and a whiteboard and pen.

Count together in fives, forwards and backwards, putting down a tower of 5 cubes each time. Practise again, starting from different multiples of five.

### Wednesday

Start by counting in fives from 0–60 and back again, using either a counting stick or a 100 square to support.

Give each pair of pupils the towers of 5 cubes from Tuesday and a whiteboard and pen.

Ask pupils to create an array from 6 towers of 5 cubes. Ask them to write both facts ($5 \times 6 = 30$ and $6 \times 5 = 30$).

Now ask how many rows of 5 there are in 30. Model writing $30 \div 5 = 6$. Next ask pupils to rotate the array to see how many rows of 6 there are in 30. Model writing $30 \div 6 = 5$. Highlight all the facts relating to this array ($30 = 6 \times 5$, $6 = 30 \div 5$, etc.).

### Thursday

Give each pair of pupils the towers of 5 cubes from Wednesday and a whiteboard and pen.

Call out a multiple of five as a multiplication table fact (e.g. $3 \times 5$). Ask partner 1 in each pair to make an array that demonstrates this. Partner 2 records both multiplication facts relating to the array. Then together they write the two related division facts.

### Friday

Repeat Thursday's activity, with pupils swapping roles.

# Week 4: Multiplication and division

## Recall and use multiplication and division facts for the 10 multiplication table

**Resources:** counting stick / 100 square, Dienes sets

**Vocabulary:** × lots of, groups of, times, multiply, multiplied by, multiple of, once, twice, three times … twelve times, *n* times as (big, long, wide, etc.), repeated addition, array, row, column, double, ÷, halve, share, share equally, one each, two each, three each …, group in pairs, threes … tens, equal groups of, divide, divided by, divided into, left, left over

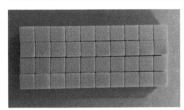

### Monday

Start by counting in tens from 0–100 and back again, using either a counting stick or a 100 square to support.

Give each pair of pupils 4 Dienes rods.

Write '4 groups of 10 are equal to 40' on the board and ask pupils to prove it in as many different ways as they can using the Dienes rods. Take photographs to display throughout the week.

### Tuesday

Start by counting in tens from 0–120 and back again, using either a counting stick or a 100 square to support. Each time you count in tens forwards and backwards, pupils put down or remove a rod.

Give each pair of pupils 12 Dienes rods.

Ask pupils to place 6 rods in an array. Ask pupils to rotate the array to see if it makes any difference. Is the array still the same shape? Encourage discussion using correct vocabulary. Repeat with other 10 multiplication table facts.

### Wednesday

Start by counting in tens from 0–120 and back again, using either a counting stick or a 100 square to support.

Give each pair of pupils 5 Dienes rods.

Ask pupils to place the 5 rods in an array and write both facts (5 × 10 = 50 and 10 × 5 = 50). Ask them how many rows of 10 there are in 50. Model writing 50 ÷ 10 = 5. Now ask pupils to rotate the array to see how many rows of 5 there are in 50. Model writing 50 ÷ 5 = 10. Discuss 5 = 50 ÷ 10, etc.

### Thursday

Give each pair of pupils 12 Dienes rods and a whiteboard and pen.

Call out a multiple of ten as a multiplication table fact (e.g. 7 × 10). Partner 1 in each pair makes an array that demonstrates this while partner 2 records both multiplication facts relating to the array. Pupils write the related division facts together.

### Friday

Repeat Thursday's activity, with pupils swapping roles.

# Week 5: Multiplication and division

## Recall and use multiplication and division facts for the 2, 5 and 10 multiplication tables

**Resources:** counting stick / 100 square, coins

**Vocabulary:** × lots of, groups of, times, multiply, multiplied by, multiple of, once, twice, three times … twelve times, *n* times as (big, long, wide, etc.), repeated addition, array, row, column, double, ÷, halve, share, share equally, one each, two each, three each …, group in pairs, threes … tens, equal groups of, divide, divided by, divided into, left, left over

### Monday

Start by counting in twos from 0–24 and back again, using either a counting stick or a 100 square to support.

Give each pair of pupils 12 × 2p coins.

Call out a 2 multiplication table fact and ask pupils to lay out that number of coins. They count in twos, pointing at each coin as they count up, and then count back to 0 in the same way. Repeat with a different multiple of two.

### Tuesday

Start by counting in fives from 0–60 and back again, using either a counting stick or a 100 square to support.

Give each pair of pupils 12 × 5p coins.

Call out a 5 multiplication table fact and ask pupils to lay out that number of coins. They count in fives, pointing at each coin as they count up. Repeat with a different multiple of five.

### Wednesday

Repeat Tuesday's activity, this time starting with a multiple of 5p and counting out that total, then counting back to zero and removing 5p each time.

### Thursday

Start by counting in tens from 0–120 and back again, using either a counting stick or a 100 square to support.

Give each pair of pupils 12 × 10p coins.

Call out a 10 multiplication table fact and ask pupils to lay out that number of coins. They count in tens, pointing at each coin as they count up. Repeat with a different multiple of ten.

### Friday

Repeat Thursday's activity, this time starting with a multiple of 10p and counting out that total, then counting back to zero and removing 10p each time.

# Week 6: Multiplication and division

## Show that multiplication of two numbers can be done in any order (commutative) and division of one number by another cannot

**Resources:** counting stick / 100 square, counters, Dienes

**Vocabulary:** × lots of, groups of, times, multiply, multiplied by, multiple of, once, twice, three times … twelve times, *n* times as (big, long, wide, etc.), repeated addition, array, row, column, double, ÷, halve, share, share equally, one each, two each, three each …, group in pairs, threes … tens, equal groups of, divide, divided by, divided into, left, left over

### Monday

Give each pair of pupils some counters and a whiteboard and pen.

Start by counting in twos from 0–24 and back again, using either a counting stick or a 100 square to support.

Call out a 2 multiplication table fact. Ask partner 1 in each pair to create the array while partner 2 writes the two multiplication facts and two related division facts.

Emphasise that multiplication can be done in any order (commutativity), whereas division cannot.

### Tuesday

Give each pair of pupils some counters and a whiteboard and pen.

Start by counting in fives from 0–60 and back again, using either a counting stick or a 100 square to support.

Call out a 5 multiplication table fact (e.g. 5 × 3). Ask partner 1 in each pair to create the array while partner 2 writes the two multiplication facts and the two related division facts.

Emphasise that multiplication can be done in any order (commutativity), whereas division cannot.

### Wednesday

Repeat Tuesday's activity, with pupils swapping roles.

### Thursday

Give each pair of pupils some Dienes and a whiteboard and pen.

Start by counting in tens from 0–100 and back again, using either a counting stick or a 100 square to support.

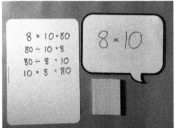

Call out a 10 multiplication table fact (e.g. 10 × 2). Ask partner 1 in each pair to create the array using Dienes while partner 2 writes the two multiplication facts and the two related division facts.

Emphasise that multiplication can be done in any order (commutativity), whereas division cannot.

### Friday

Repeat Thursday's activity, with pupils swapping roles.

# Week 1: Fractions

## Recognise, find, name and write fractions $\frac{1}{2}$ of a length, shape, set of objects or quantity

**Resources:** counting stick, objects

**Vocabulary:** part, unit, equal parts, fraction, share, share equally, group in pairs, twos, fours …, equal groups of, divide, divided by, divided into, left, left over, one whole, one half, two halves, one quarter … four quarters, numerator, denominator

### Monday

Give each pupil 12 objects.

Start by counting in halves from 0–10 and back again. If needed, use a number line to support pupils during the week.

Ask pupils to count out 8 objects and then to halve the set by sharing them equally into two groups. Ask how many there are in each half. Explain that it is important that all parts are equal when we work with fractions. Repeat, halving 10 and then 12 objects.

### Tuesday

Give each pupil 24 objects.

Start by counting in halves from 0–10 and back again.

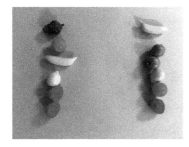

Ask pupils to count out 12 objects and then to halve the set by sharing them equally into two groups. Ask how many there are in each half. Explain that it is important that all parts are equal when we work with fractions. Repeat, halving 14, 16, 18, 20, 22 and 24 objects.

### Wednesday

Start by counting in halves from 0–10 and back again.

Set out different quantities of objects and ask pupils to work in pairs to find half of the set of objects, as per Tuesday's activity.

### Thursday

Explain that today pupils will be learning how to read and write fractions.

Write $\frac{1}{2}$ on the board. Explain that the bottom number of the fraction is called the denominator and tells us how many parts the whole / unit has been split equally into (2). The top number is called the numerator and tells us how many of those equal parts have been taken (1).

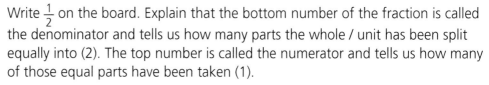

Give each pupil 10 objects. Ask them to halve the set by sharing them equally into two groups. Ask how many objects there are in one of the equal parts. Write the number sentence '$\frac{1}{2}$ of 10 = 5'.

### Friday

Repeat Thursday's activity, finding half of various sets of objects and writing the number sentence.

**Recognise, find, name and write fractions $\frac{1}{4}$ of a length, shape, set of objects or quantity**

**Resources:** counting stick, strips of paper, objects, paper shapes, cubes

**Vocabulary:** part, unit, equal parts, fraction, share, share equally, group in pairs, twos, fours …, equal groups of, divide, divided by, divided into, left, left over, one whole, one half, two halves, one quarter … four quarters, numerator, denominator

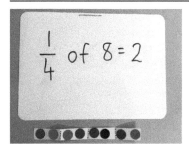

### Monday

Start by counting in quarters from $0-2\frac{1}{2}$ and back again. If needed, use a counting stick to support during the week. Give each pupil a strip of paper and 8 objects.

Pupils fold their strip into four equal sections (by halving, then halving again). Explain that they have just split one whole piece of paper into four equal parts (quarters). Ask them to share the 8 objects equally by placing them on the sections. Ask how many objects there are in one of the quarters. Write '$\frac{1}{4}$ of 8 = 2' on the board.

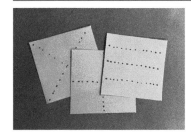

### Tuesday

Start by counting in quarters from $0-2\frac{1}{2}$ and back again. Give each pupil paper squares, circles and rectangles and a whiteboard and pen.

Ask pupils to explore as many ways as possible to divide the square into quarters. Once pupils are happy that they have found a way of quartering the shape, they should draw it on their whiteboard. If time permits, pupils could cut out one quarter of each square and discuss that each quarter, although a different shape, has the same area. Repeat with circles and rectangles.

### Wednesday

Start by counting in quarters from $0-2\frac{1}{2}$ and back again. Give each pupil 20 objects and a whiteboard and pen.

Ask pupils to count out 12 objects and divide the set by sharing them equally into four groups. Ask how many objects are in each quarter. Explain that it is important that all parts are equal when we work with fractions. Repeat, quartering 16 and then 20 objects.

### Thursday

Start by counting in quarters from $0-2\frac{1}{2}$ and back again.

Repeat Wednesday's activity, giving each pupil or pair 24 cubes. They find one quarter of 12, 16, 20 and 24 and then write the number sentences.

### Friday

Show a length of 4 cubes. Ask pupils to show you one quarter of it. Repeat for 8, 12 and 16 cubes.

## Recognise, find, name and write fractions $\frac{2}{4}$ of a length, shape, set of objects or quantity

**Resources:** counting stick, strips of paper, objects, cubes

> **Vocabulary:** part, unit, equal parts, fraction, share, share equally, group in pairs, twos, fours …, equal groups of, divide, divided by, divided into, left, left over, one whole, one half, two halves, one quarter … four quarters, numerator, denominator

### Monday

Start by counting in quarters from $0-2\frac{1}{2}$ and back again. Stop at one half and highlight the relationship between one half and two quarters. If needed, use a counting stick to support pupils during the week.

Write $\frac{1}{4}$ on the board. Explain that the bottom number (denominator) tells us how many equal parts the whole/unit has been split into (4). The top number (numerator) tells us how many equal parts have been taken (1). Write '$\frac{2}{4}$ of 4 = 2' on the board. Repeat, finding $\frac{2}{4}$ of 8 and then of 12 objects.

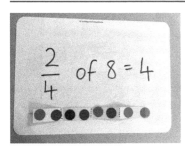

### Tuesday

Start by counting in quarters from $0-2\frac{1}{2}$ and back again. Give each pupil a strip of paper and 8 objects.

Pupils fold their strips into four equal sections (by halving, then halving again). Explain that they have just split one whole piece of paper into four equal parts (quarters). They then share the 8 objects equally by placing them on the sections. Ask how many counters there are in two of the quarters. Write '$\frac{2}{4}$ of 8 = 4' on the board. If time permits, try $\frac{1}{2}$ of 8 to compare and ascertain that $\frac{1}{2}$ is the same as $\frac{2}{4}$.

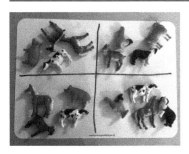

### Wednesday

Start by counting in quarters from $0-2\frac{1}{2}$ and back again. Give each pupil 20 objects.

Ask pupils to count out 12 objects and divide the set by sharing them equally into four groups. Ask how many objects there are in two quarters. Explain that it is important that all parts are equal when we work with fractions. Repeat, finding two quarters of 16 and then 20 objects.

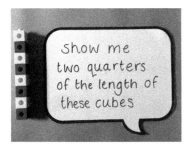

### Thursday

Start by counting in quarters from $0-2\frac{1}{2}$ and back again.

Show a length of 4 cubes. Ask pupils to show you one quarter, then two quarters of it. Repeat for 8, 12 and 16 cubes.

### Friday

Repeat Thursday's activity, starting with a length of 20 cubes and asking pupils to show you two quarters of it. Repeat for 8, 12 and 16 cubes.

## Recognise, find, name and write fractions $\frac{3}{4}$ of a length, shape, set of objects or quantity

**Resources:** counting stick, strips of paper, objects, cubes

**Vocabulary:** part, unit, equal parts, fraction, share, share equally, group in pairs, twos, fours …, equal groups of, divide, divided by, divided into, left, left over, one whole, one half, two halves, one quarter … four quarters, numerator, denominator

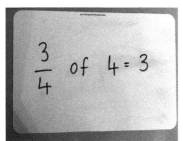

### Monday

Start by counting in quarters from $0$–$2\frac{1}{2}$ and back again. If needed, use a counting stick to support pupils during the week.

Write $\frac{1}{4}$ on the board. Remind the class that the bottom number (denominator) tells us how many equal parts the whole/the unit has been split equally into (4). The top number (numerator) tells us how many equal parts have been taken (1). Use cubes to model finding $\frac{3}{4}$ of 4 and write '$\frac{3}{4}$ of 4 = 3' on the board. Repeat for three quarters of 8 and 12 objects.

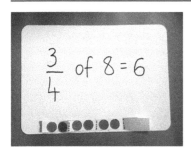

### Tuesday

Start by counting in quarters from $0$–$2\frac{1}{2}$ and back again. Give each pupil a strip of paper and 8 objects.

Pupils fold their strips into 4 equal pieces (by halving, then halving again). Explain that they have just split one whole into four equal parts (quarters). Pupils share the 8 objects equally by placing them on the sections. Ask how many there are in three of the quarters. Write '³/₄ of 8 = 6' on the board.

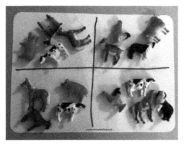

### Wednesday

Start by counting in quarters from $0$–$2\frac{1}{2}$ and back again. Give each pupil 20 objects.

Ask pupils to count out 12 cubes and divide the set by sharing them equally into four groups. Ask how many objects there are in three quarters. Explain that it is important that all parts are equal when we work with fractions. Repeat, finding three quarters of 16 and then 20 objects.

### Thursday

Start by counting in quarters from $0$–$2\frac{1}{2}$ and back again.

Show a length of 4 cubes. Ask pupils to show you one quarter, then two quarters and then three quarters. Show a stick of 8 cubes, and ask pupils to show you one quarter, then two quarters, then three quarters of the length of this stick of cubes. Repeat for three quarters of 12 and 16 cubes.

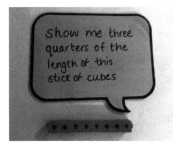

### Friday

Show a length of 20 cubes. Ask pupils to show you three quarters of it. Repeat for 8, 12 and 16 cubes.

# Week 5: Fractions

## Recognise, find, name and write fractions $\frac{1}{3}$ of a length, shape, set of objects or quantity

**Resources:** counting stick, strips of paper, objects, cubes

**Vocabulary:** part, unit, equal parts, fraction, share, share equally, group in pairs, twos, fours …, equal groups of, divide, divided by, divided into, left, left over, one whole, one half, two halves, one quarter … four quarters, numerator, denominator

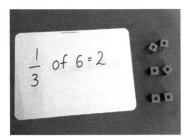

### Monday

Start by counting in thirds from 0–3 and back again. If needed, use a counting stick to support pupils during the week.

Write $\frac{1}{3}$ on the board. Remind the class that the bottom number (denominator) tells us how many equal parts the whole / unit has been split into (3). The top number (numerator) tells us how many equal parts have been taken (1). Use cubes to model finding $\frac{1}{3}$ of 6 and write '$\frac{1}{3}$ of 6 = 2' on the board. Repeat for one third of 9 and 12.

### Tuesday

Start by counting in thirds from 0–3 and back again. Give each pupil a strip of paper and 6 objects.

Pupils fold their strip into three equal sections. Explain that they have just split one whole into three equal parts (thirds). Pupils share the 6 objects equally by placing them on the sections. Ask how many counters there are in one third. Write '$\frac{1}{3}$ of 6 = 2' on the board.

### Wednesday

Start by counting in thirds from 0–3 and back again. Give each pupil 18 objects.

Pupils count out 12 objects and divide the set by sharing them equally into three groups. Ask how many objects there are in one third. Explain that it is important that all parts are equal when we work with fractions. Repeat, finding one third of 18 objects.

### Thursday

Start by counting in thirds from 0–3 and back again.

Show a length of 3 cubes. Ask pupils to show you one third of it. Repeat for one third of 6, 9 and 12 cubes.

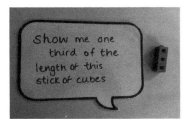

### Friday

Show a length of 15 cubes. Ask pupils to show you one third of it. Repeat for one third of 18 cubes.

# Week 6: Fractions

## Recognise, find, name and write fractions $\frac{1}{3}$, $\frac{1}{4}$, $\frac{2}{4}$ and $\frac{3}{4}$

**Resources:** counting stick, strips of paper, cubes

**Vocabulary:** part, unit, equal parts, fraction, share, share equally, group in pairs, twos, fours …, equal groups of, divide, divided by, divided into, left, left over, one whole, one half, two halves, one quarter … four quarters, numerator, denominator

### Monday

Start by counting in quarters from $0-2\frac{1}{2}$ and back again. If needed, use a counting stick to support pupils during the week.

Write $\frac{1}{4}$ on the board. Remind the class that the bottom number (denominator) tells us how many equal parts the whole / unit has been split into. The top number tells us how many equal parts have been taken. Model using dots on the strips to find one quarter of a quantity. Repeat to find $\frac{1}{4}$ of different quantities.

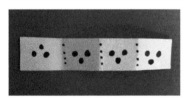

### Tuesday

Start by counting in quarters from $0-2\frac{1}{2}$ and back again.

Write $\frac{2}{4}$ on the board. Remind the class of the vocabulary for numerator and denominator. Model using dots on the strips to find two quarters of a quantity. Repeat to find $\frac{2}{4}$ of different quantities.

### Wednesday

Start by counting in quarters from $0-2\frac{1}{2}$ and back again.

Write $\frac{3}{4}$ on the board. Remind the class of the vocabulary for numerator and denominator. Model using dots on the strips to find three quarters of a quantity. Repeat to find $\frac{3}{4}$ of different quantities.

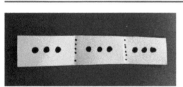

### Thursday

Start by counting in thirds from $0-2\frac{1}{3}$ and back again.

Write $\frac{1}{3}$ on the board. Remind the class of the vocabulary for numerator and denominator. Model using dots on the strips to find one third of a quantity. Repeat to find $\frac{1}{3}$ of different quantities.

show me three quarters of the length of this stick of cubes

### Friday

Show a length of 12 cubes. Ask pupils to show one half of it. Repeat for one quarter, two quarters, three quarters and one third of 12 cubes. If confident, pupils could work in pairs to explore fractions of different quantities of cube towers (e.g. a tower with 16 cubes, 18 cubes, 20 cubes).

# Blank vocabulary cards

# Notes